ANIMALS AROUND THE WORLD
All About Antarctic
Humpback Whales
EZ READERS
An Imprint of Mitchell Lane
Meg Greve

Creating Young Nonfiction Readers

EZ Readers lets children delve into nonfiction at beginning reading levels. Young readers are introduced to new concepts, facts, ideas, and vocabulary.

Tips for Reading Nonfiction with Beginning Readers

Talk about Nonfiction

Begin by explaining that nonfiction books give us information that is true. The book will be organized around a specific topic or idea, and we may learn new facts through reading.

Look at the Parts

Most nonfiction books have helpful features. Our *EZ Readers* include color photographs and graphic aids, a table of contents, a glossary, and an index. Share the purpose of these features with your reader.

Color Photos and Graphic Aids

A lot of information can be found by "reading" photos, charts, maps, and other graphic aids found within nonfiction texts. Help your reader learn more about the different ways information can be displayed.

Table of Contents

Located at the front of the book, this list shows the big ideas within the text and the page numbers where they can be found.

Glossary

Located at the back of the book, the glossary defines key words and phrases that are related to the topic. These words and phrases can be found in the text in colored type.

Index

Located at the back of the book, an index is an alphabetical list of topics and the page numbers where they can be found.

With a little help and guidance about reading nonfiction, you can feel good about introducing a young reader to the world of *EZ Readers* nonfiction books.

Mitchell Lane
PUBLISHERS

2001 SW 31st Avenue
Hallandale, FL 33009
mitchelllanepub.com

First Edition, 2026.

Author: Meg Greve
Designer: Rhea Magaro
Editor: Kim Thompson

Names/credits:
Title: All about Antarctic Humpback Whales / by Meg Greve
Description: Hallandale, FL :
Mitchell Lane Publishers, [2026]

Series: Animals Around the World
Library bound ISBN: 979-8-89260-516-8
eBook ISBN: 979-8-89260-525-0

Library of Congress Control Number: 2025933681

EZ Readers is an imprint of
Mitchell Lane Publishers

PHOTO CREDITS
Shutterstock: Philipp Salveter, cover, 1; Mathias Berlin, 5; Tomas Kotouc, 7; Manuel Balesteri, 8; Stephen Allen Photography, 10; Danita Delimont, 12; James C Farr, 14; Paul S. Wolf, 16; martin_hristov, 19; Kertu, 21; deer boy, 22.

Contents

The Southern Ocean surrounds the continent of Antarctica. It is full of glaciers and icebergs. But it is perfect for the humpback whale! Humpbacks swim, hunt, and dive here from November through March.

Humpback whales can be more than 43 feet (13 meters) long. That is longer than a school bus! Their fins and tail **flukes** help them dive deep in the water or leap high into the air. **Blubber** under their skin keeps them warm.

Humpbacks have bumps all over their heads. These golf ball-sized lumps are called tubercles. Each one has a hair growing out of it. Tubercles sense movement in the water. They help the whale know when food is close.

Humpback whales are **baleen** whales. They do not have teeth. Instead, they have bristles in their mouths for water to flow through. The bristles filter out **krill**, fish, and other small animals. Then, the whale swallows a tasty meal.

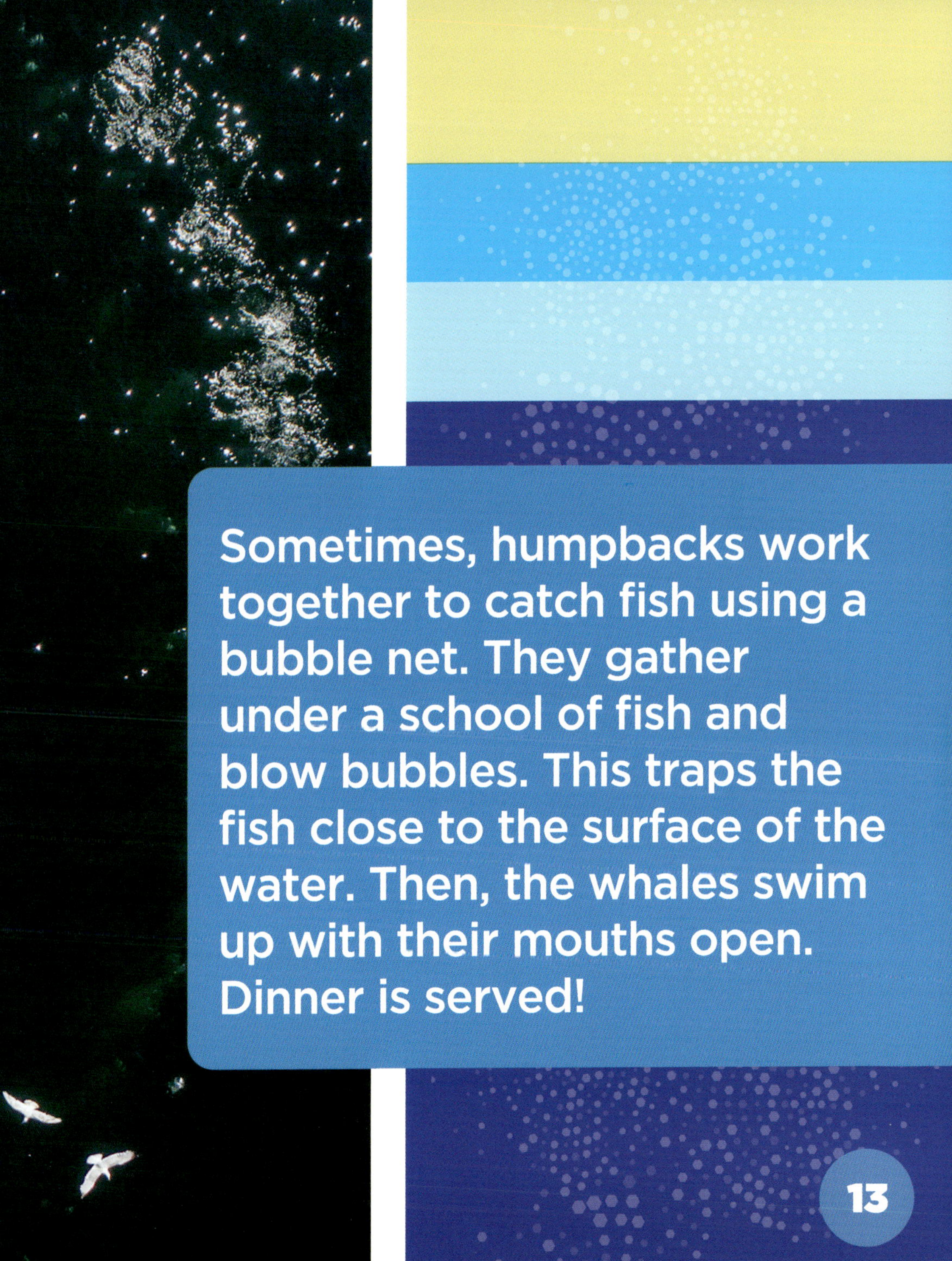

Sometimes, humpbacks work together to catch fish using a bubble net. They gather under a school of fish and blow bubbles. This traps the fish close to the surface of the water. Then, the whales swim up with their mouths open. Dinner is served!

Humpback whales live alone or in small groups. They are known for the songs they sing. Scientists think that is how they talk to each other.

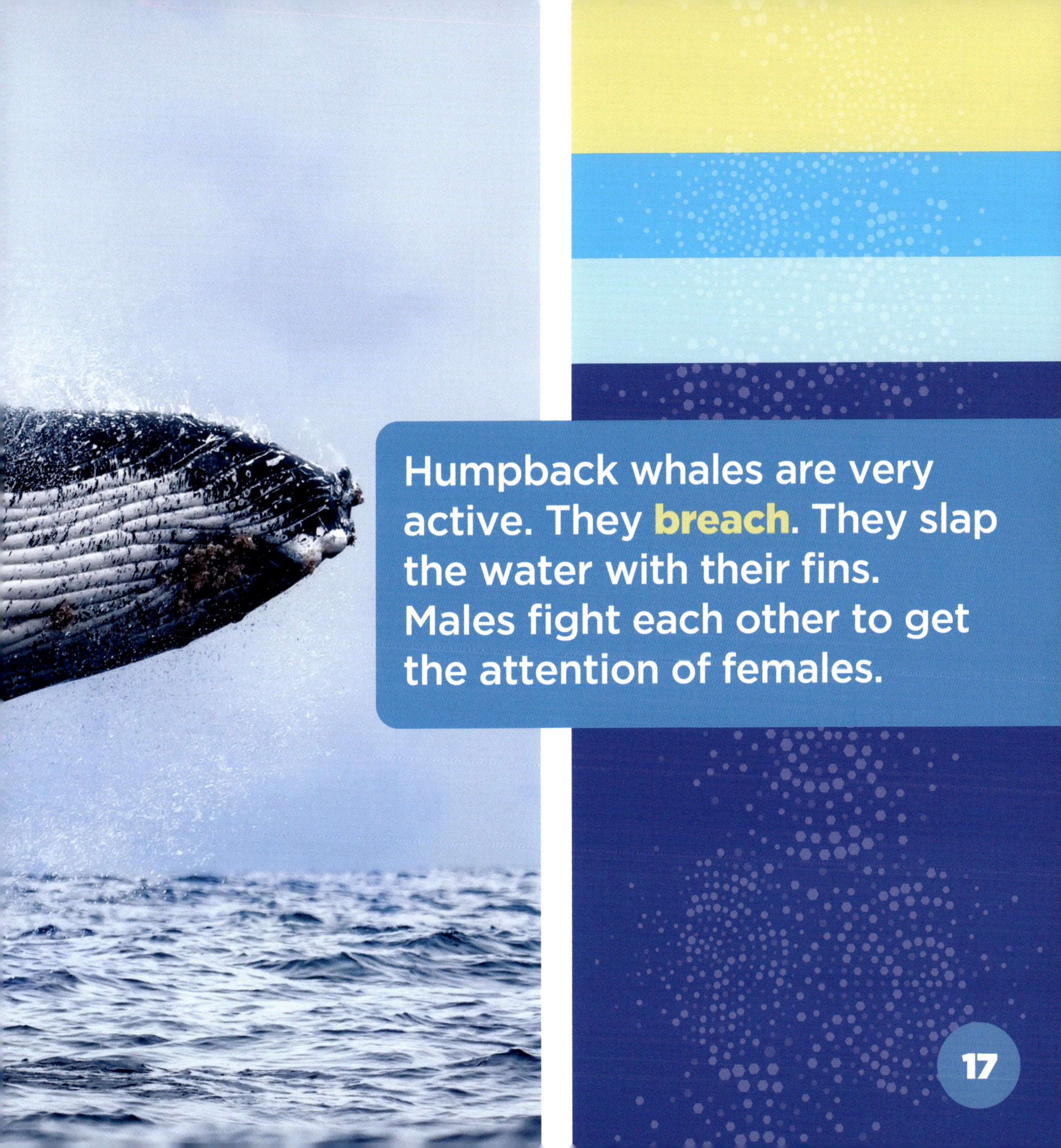

Humpback whales are very active. They **breach**. They slap the water with their fins. Males fight each other to get the attention of females.

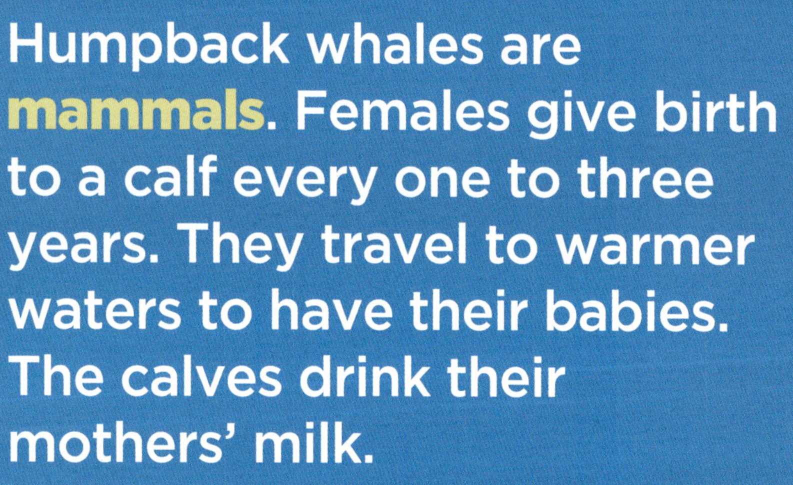

Humpback whales are **mammals**. Females give birth to a calf every one to three years. They travel to warmer waters to have their babies. The calves drink their mothers' milk.

Babies learn to dive, breach, and hunt from their mothers. When they are about one year old, calves leave to live on their own and start their own families.

Where Do Humpback Whales Live?

Humpback whales can be found in every ocean around the world. Many spend the summer near Antarctica and the winter in warmer waters. Remember that in Antarctica, it is summer from October to February.

Interesting Facts

- Humpback whales get their name from the hump below their dorsal (back) fin.
- Like a fingerprint, each fluke on a humpback has its own markings and patterns.
- Calves and mothers touch fins to show affection for each other.
- Humpback whales have two blowholes they use to breathe when they come up for air.

Parts of a Humpback Whale

baleen
Bristle-like baleen plates are made of the same material as fingernails. They act like sieves to filter water and capture food.

blowholes
Blowholes are used for breathing. Humpbacks do not breathe through their mouths.

fins
Fins help humpbacks steer and swim.

fluke
Each fluke is one half of a humpback's double-sided tail.

throat grooves
These places stretch so that the whale's mouth can take in lots of water while feeding.

tubercles
The tubercles each have one hair that senses movement under the water.

Glossary

baleen (buh-LEEN)
Plates made of the same material as fingernails that are found in the jaws of toothless whales

blubber (BLUHB-ur)
A layer of fat under the skin of an animal that lives in a cold place

breach (breech)
To jump out of the water and slam back to the surface of the water

flukes (flooks)
The left and right lobes of a whale's tail

krill (kril)
Small, shrimp-like creatures that are an important source of food for many ocean animals

mammals (MAM-uhlz)
Warm-blooded animals that have backbones and fur or hair, that give birth to live babies, and that feed their babies milk from the mother

Index

Further Reading

deVos, Asha. *Humpback Whale: A First Field Guide to the Singing Giant of the Ocean.* Macmillan, 2022.

Gleisner, Jenna Lee. *Humpback Whales.* Jump!, 2023.

On the Internet

Active Wild: Antarctica Facts for Kids
www.activewild.com/antarctica-facts-for-kids
Learn all about Antarctica and the animals that live there.

National Geographic: Humpback Whale
www.nationalgeographic.com/animals/mammals/facts/humpback-whale
Learn amazing facts about humpback whales.